HEREDITARY
CHARACTER
and TALENT

FRANCIS GALTON

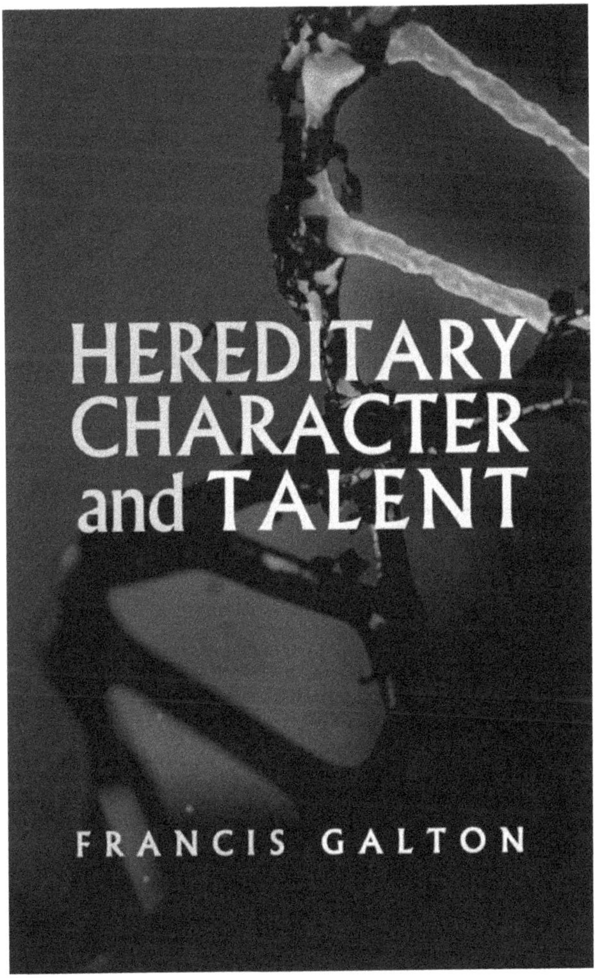

HEREDITARY CHARACTER and TALENT

FRANCIS GALTON

Reprinted by

Hereditary Character and Talent
by Francis Galton

ISBN 978-1-947844-78-0

Reprinted by Suzeteo Enterprises, 2018. All Rights Reserved.

Part one originally appeared in:
Macmillan's Magazine, vol. 12, 1865 pp. 157-166

Part two originally appeared in:
MacMillan's Magazine, vol. 12, 1865 pp. 318-327.

A note from the publisher: this edition was reproduced from exact facsimiles of the original. Anglicisms have been retained. Galton's use of fractions was sometime converted into word form, but this obviously does not change the meaning of the content.

Galton included two tables in his articles. These were carefully re-created. However, since these tables were hard to read in some places, errors may exist within the tables. Reproductions of the facsimiles for these tables are appended to the end of this edition for the reader to consult in case of questions or ambiguities.

Hereditary Character and Talent

Part I[1]

The power of man over animal life, in producing whatever varieties of form he pleases, is enormously great. It would seem as though the physical structure of future generations was almost as plastic as clay, under the control of the breeder's will. It is my desire to show, more pointedly than — so far as I am aware— has been attempted before, that mental qualities are equally under control.

A remarkable misapprehension appears to be current as to the fact of the transmission of talent by inheritance. It is commonly asserted that the children of eminent men are stupid; that, where great power of intellect seems to have been inherited, it has descended through the mother's side; and that one son commonly runs away with the talent of a whole family. My own inquiries have led me to a diametrically opposite conclusion. I find that talent is transmitted by inheritance in a very remarkable degree; that the mother has by no means the monopoly of its transmission; and that whole families of persons of talent are more common than those in which one member only is possessed of it. I justify my conclusions by the statistics I now proceed to adduce, which I believe are amply sufficient to command conviction. They are only a part of much material I have collected, for a future volume on this subject; all of which points in the same direction. I

---

[1] *Macmillan's Magazine*, vol. 12, 1865 pp. 157-166

should be very grateful to any of my readers for information that may help me in my further inquiries.

In investigating the hereditary transmission of talent, we must ever bear in mind our ignorance of the laws which govern the inheritance even of physical features. We know to a certainty that the latter exist, though we do not thoroughly understand their action. The breeders of our domestic animals have discovered many rules by experience, and act upon them to a nicety. But we have not advanced, even to this limited extent, in respect to the human race. It has been nobody's business to study them; and the study is difficult, for many reasons. Thus, only two generations are likely to be born during the life of any observer; clothing conceals the shape; and each individual rarely marries more than once. Nevertheless, all analogy assures us that the physical features of man are equally transmissible with those of brutes! The resemblances between parent and offspring, as they appear to a casual observer, are just as close in one case as in the other; and, therefore, as a nearer scrutiny has established strict laws of hereditary transmission in brutes, we have every reason for believing that the same could also be discovered in the case of man.

So far as I am aware, no animals have ever been bred for general intelligence. Special aptitudes are thoroughly controlled by the breeder. He breeds dogs that point, retrieve, that fondle, or that bite; but no one has ever yet attempted to breed for high general intellect, irrespective of all other qualities. It would be a most interesting subject for an attempt. We hear constantly of prodigies of dogs, whose very intelligence makes them of little value as slaves. When they are wanted, they are apt to be absent on their own errands. They are too critical of their master's conduct. For instance, an intelligent dog shows marked contempt for an unsuccessful sportsman. He will follow nobody along a road that leads on a well-known tedious errand. He does not readily forgive a man who wounds his self-esteem. He is often a dexterous thief and a sad hypocrite. For these reasons an over-intelligent dog is not an object of particular desire, and therefore, I suppose, no one has ever thought of encouraging a breed of wise dogs. But it would be a most interesting occupation for a country philosopher to pick up the cleverest dogs he could hear of, and mate them together, generation after generation — breeding purely for intellectual power, and disregarding shape, size, and every other quality.

As no experiment of this description has ever been made, I cannot appeal to its success. I can only say that the general resemblances in mental qualities between parents and offspring, in man and brute, are every whit as near as the resemblance of their physical features; and I must leave the existence of actual laws in the former case to be a matter of inference from the analogy of the latter. Resemblance frequently fails where we might have expected it to hold; but we may fairly ascribe the failure to the influence of conditions that we do not yet comprehend. So long as we have a plenitude of evidence in favour of the hypothesis of the hereditary descent of talent, we need not be disconcerted when negative evidence is brought against us. We must reply that just the same argument might have been urged against the transmission of the physical features of our domestic animals; yet our breeders have discovered certain rules, and make their living by acting upon them. They know, with accurate prevision, when particular types of animals are mated together, what will be the character of the offspring. They can say that such and such qualities will be reproduced to a certainty. That others are doubtful; for they may appear in some of the descendants and not in the rest. Lastly, that there are yet other qualities, excessive in one parent an defective in the other, that will be counterbalanced and be transmitted to the offspring in a moderate proportion.

I maintain by analogy that this prevision could be equally attained in respect to the mental qualities, though I cannot prove it. All I can show is that talent and peculiarities of character are found in the children, when they have existed in either of the parents, to an extent beyond all question greater than in the children of ordinary persons. It is a fact, neither to be denied nor to be considered of importance, that the children of men of genius are frequently of mediocre intellect. The qualities of each individual are due to the combined influence of his two parents; and the remarkable qualities of the one may have been neutralized in the offspring, by the opposite or defective qualities of the other. It is natural hat contrast of qualities, in the parents' dispositions, should occur as frequently as harmony; for one of the many foundations of friendship and of the marriage union is a difference of character; each individual seeking thereby to supplement the qualities in which he feels his own nature to be deficient. We have also good reason to believe that every special talent or character depends on a variety of obscure conditions, the analysis of which has never yet been seriously attempted. It is easy to conceive that the entire character might be considerably altered, owing to the modification of any one of these conditions.

As a first step in my investigation, I sought a biographical work, of manageable size, that should contain the lives of the chief men of genius whom the world is known to have produced. I ultimately selected that of Sir Thomas Phillips, in his well-known work of reference, "The Million of Facts" because it is compiled with evident discrimination, and without the slightest regard to the question on which I was engaged. It is, moreover, prefaced, — "It has been attempted to record, in brief, only the ORIGINAL MINDS, who founded or originated. Biography in general is filled with mere imitators, or with men noted only for chance of birth, or necessary position in society." I do not mean to say that Sir Thomas Phillips's selection is the best that could have been made, for he was a somewhat crotchety writer. It did not, however, much matter whose biography I adopted, so long as it had been written in the abovementioned spirit, and so long as I determined to abide steadfastly within its limits, without yielding to the temptation of supplying obvious omissions, in a way favourable to any provisional theory.

According to this select biography, I find that 605 notabilities lived between the years 1453 and 1853. And among these are no less than 102 relationships, or 1 in 6, according to the following list: —

Art	Lit. & science	More distant.	Brothers	Father & son.	Number.	Notable Persons.
	1			2	3	J. Adams, Pres. U.S.A.; son Samuel also patriot; nephew, J. Quincy, president.
	2		2		2	W. Belsham, historian; brother of T. Belsham, Unitarian minister.
	3			2	3	J. Bernouilli, father of James and uncle of John, all mathematicians.
3				2	3	Breughel, father and two sons, painters.
	2			2		Buxtorff, father and son, Hebraists.
3	1		2		3	Caracci, An. and Ag. brothers, Lud. cousin, painter.
	1		2		2	Cartwright, reformer; brother, mechanist.
	3			3	3	Casini, grandfather, and son, all mathematicians.
	1	2			2	Cooper, Privy Councillor to Cromwell; grandson, literary.
			2		2	De Witt, two brothers, patriots.
				3	3	Elizabeth, queen, daughter of Henry VIII. and granddaughter of Sir. T. Bullen.
	2	2			2	Fontana, two brothers natural philosophers.
	2			2	2	Forster, father and son, naturalists (Cook's voyages).
	6			6	6	Gronovious, sons and grandsons, six in all, learned critics.
		1		2	3	Gustavus Adolphus, father of Christina and grandson of Gustavus Vasa.
	2			2	2	Herschel, father and son, astronomers.
	2		2		2	Hunter, two brothers, anataomists.
	2	2			2	Jussieu, uncle and nephew, botanists.
				4	4	Medici, grandfather, father, and son, and Catherine.
				2	2	Orleans, Egalité, and son Louis Phillippe.
			2		2	Ostade, two brothers, painters.
4			4		4	Perrault, four brothers, all writers.
	1			2	2	Penn, admiral; son, Quaker writers.
		2		2	4	Phillibert, Prince of Orange; cousin William, whose son was Maurice. His grandson was our William III.
				2	2	Pitt, father and son, statesmen.
	2			2	2	Scaliger, classical critic; son also.
				2	2	Sforzas, father and son.
	1	2			2	Shaftesbury, statesman; grandson, author.
	1			2	2	Sheridan, father and son.
	1			2	2	Staël, Madam, daughter of Necker, financier.
	6			6	6	Stephens, family of six, critics and editors.
2				2	2	Teniers, father and son, painters.
				2	2	Tytler, historian and poet; son, Lord Woodhouselee.
2				2	2	Vanderwelde, father and son, painters.
	2		2		2	Vanderwurf, two brothers, famous for small history.
		1	2		3	Valnoo, two brothers, and nephew, painters.
	1			2	2	Walpole, Sir Robert, statesman; Sir Horace, author.
				2	2	Van Tromp, father and son, admirals.
	1	2			2	Villiers, statesman; grandson, the reprobate poet.
	2			2	2	Vossius, father, son, and other relatives, all writers.
	2			2	2	Warton, editor of Pope; son, poet.

9 | 52 | 14 | 22 | 66 | 102

It will be observed[2] that the number is swelled by four large families, such as those of Gronovius and Stephens, of six members each, and of the Medici and the House of Orange, of four members each. The two first might be objected to, as hardly worthy of the distinguished place they occupy. But we must adhere to our biography; there are many more relationships that could very fairly have been added, as a set-off against these names. Such are two more Vanderweldes, and the family of Richelieu; besides others, like Hallam the historian, and Watt the mechanic, whose sons died early, full of the highest promise. Even if sixteen names were struck out of our list, the proportion of the relationship would remain as

86/605 , or 1 in 7. And these are almost wholly referable to transmission of talent through the male line; for eminent mothers do not find a place in mere biographical lists. The overwhelming force of a statistical fact like this renders counter-arguments of no substantial effect.

Next, let us examine a biographical list of much greater extension. I have selected for this purpose an excellent brief dictionary by Mr C. Hone. It is not yet published, but part of its proof sheets have been obligingly lent to me. The entire work appears to contain some 19,000 names; it is, therefore, more than thirty times as extensive as the list we have hitherto been considering. I have selected one part only of this long series of names for examination, namely, those that begin with the letter M. There are 1141 names that remain under this letter, after eliminating those of

---

[2] Editor: A facsimile of this chart is appended to the end of this edition.

sovereigns, and also of all persons who died before A.D. 1453. Out of these, 103, or 1 in 11, are either fathers and sons, or brothers; and I am by no means sure that I have succeeded in hunting out all the relationships that might be found to exist among them.

It will be remarked that the proportion of distinguished relationships becomes smaller, as we relax the restrictions of our selection; and it is reasonable that it should be so, for we then include in our lists the names of men who have been inducted into history through other conditions than the possession of eminent talent.

Again, if we examine into the relationships of the notabilities of the present day, we obtain even larger proportions. Walford's "Men of the Time" contains an account of the distinguished men of England, the Continent, and America, who are now alive. Under the letter A there are 85 names of men, and no less than 25 of these, or 1 in three and a half, have relatives also in the list; 12 of them are brothers, and 11 fathers and sons.

Abbott, Rev. Jacob (U.S.A.), author on religious and moral subjects.
Abbott, Rev. John, younger *brother* of above, author on religious and moral subjects.
A'Beckett, Sir William, author, Solicitor-Gen. of New South Wales, and *brother* of late Gilbert Abbott A'Beckett.
Adam Jean Victor, painter, *son* of an eminent engraver.
Adams, American minister, *son* of John Quincey Adams.
Ainsworth, William Francis, editor of 'Journal of natural and Geographical Science, "Explorations in Asia Minor

and Kurdistan."

Ainsworth, William Harrison, novelist, *cousin* of above.

Aïvazooski, Gabriel, Armenian, born in the Crimea, Professor of European and Oriental languages, and member of Historical Institute of France.

Aïvazooski, Ivan, a marine painter, *brother* of above.

Albermarle, Earl of (*brother* Kepel).

Albert, Prince (*brother*).

Aldis, Sir Charles, medical.

Aldis, Charles J.B. medical, *son* of above.

Alexander, James Waddell, American divine (*son* of a Professor).

Alexander, Joseph Addison, Professor of Ancient languages, and of Biblical and Ecclesiastical history, *brother* of the above.

Alison, Sir Archibald, historian, *son* of author of "Essays on Taste:" his mother belonged to "a family which has for two centuries been eminent in mathematics and the exact sciences."

Ampère, member of French Academy, and Professor in College of France (literary), *son* of the celebrated physicist of the same name.

Arago, Etienne, journalist and theatrical writer, *brother* of the celebrated philosopher.

Argyropopulo, statesman, *son* of grand interpreter to the Porte.

Aristarchi, ecclesiastic and statesman, *son* of grand interpreter to the Porte.

Arnold, Matthew, *son* of late Dr Arnold.

Arwidson, Librarian R. Library, Stockholm, author, *son* of a person who held a high position in the Church.

Ashburton, Lord, *son* of Rt Hon. Alexander Baring.

Azeglio, Massimo, statesman and painter.

Azeglio, Marquis, *nephew* of above, diplomatist and painter.

So if we examine the biographies of artists. In Bryan's large "Dictionary of Painters," the letter A contains 391 names of men, of whom 65 are near relatives, or 1 in 6: 33 of them are fathers and sons, 30 are brothers. In Fétis "Biographie Universelle des Musiciens" the letter A contains 515 names, of which 50 are near relations, or 1 in 10. Two-third are fathers and sons, one-third are brothers.

It is justly to be urged, in limitation of the enormous effect of hereditary influence, implied by the above figures, that when a parent has achieved great eminence, his son will be placed in a more favourable position for advancement, than if he had been the son of an ordinary person. Social position is an especially important aid to success in statesmanship and generalship; for it is notorious that neither the Legislature nor the army afford, in their highest ranks, an open arena to the ablest intellects. The sons of the favoured classes are introduced early in life to both these fields of trial, with every encouragement to support them. Those of the lower classes are delayed and discouraged in their start; and when they are near the coveted goal, they find themselves aged. They are too late: they are not beaten by the superior merit of their contemporaries, but by time; as was once touchingly remarked by Sir De Lacy Evans.

In order to test the value of hereditary influence with greater precision, we should therefore extract from our biographical list the names (they are 330) of those that have achieved distinction in the more open fields of science and literature. There is no favour here beyond the advantage of a good education.

Whatever spur may be given by the desire to maintain the family fame, and whatever opportunities are afforded by abundant leisure, are more than neutralised by those influences which commonly lead the heirs of fortune to idleness and dilettantism.

Recurring to our list, we find fifty-one literary men who have distinguished relations. Therefore, no less than 51/605, or one distinguished man in every twelve, has a father, son, or brother, distinguished in literature. To take a round number at a venture, we may be sure that there have been far more than a million students educated in Europe during the last four centuries, being an average of only 2500 in each a year. According to our list, about 330 of these, or only 1 in 3000, achieved eminent distinction: yet of those who did so, 1 in 12 was related to a distinguished man. Keeping to literature alone, it is 51 to 330=1 to six and a half, that a very distinguished literary man has a very distinguished literary relative, and it is (leaving out the Gronovius and Stephenses) 20 to 330=1 to 16, and 12 to 330=1 to 28, that the relationship is father and son, or brother and brother, respectively.

The Law is, by far, the most open to fair competition of all the professions; and of all offices in the law there is none that is more surely the reward of the most distinguished intellectual capacity than that of the Lord Chancellor. It therefore becomes an exceedingly interesting question to learn what have been the relationships of our Lord Chancellors. Are they to any notable degree the children, or the parents, or the brothers of very eminent men? Lord Campbell's "Lives of the Chancellors" forms a valuable biographical dictionary for the purpose of this investigation. I have taken it just as it stands; including, as Lord Campbell does, certain Lord Keepers and Commissioners of the Great Seal, as of equal rank with the Chancellors. I may further mention, that many expressions in Lord Campbell's works show that he was a disbeliever in hereditary influence.

Now what are the facts? Since Henry VIII's time, when Chancellors ceased to be ecclesiastics, and were capable of marrying, we have had thirty-nine Chancellors, &c. whose lives have been written by Lord Campbell, or whom the following had eminent relationships:—

Sir Nicholas Bacon, Lord Keeper: son, Lord Chancellor Bacon.
Coventry: son of a very learned judge of the Common Pleas.
Bacon: father as above.
Littleton: son of a judge.
Whitelock: son of a judge, father of two sons, one of great eminence as a lawyer, the other as a soldier.
Herbert: three sons. One had high command in army; the second, the great naval officer, created Lord Torrington; the third, Chief Justice of Queen's Bench.
Finch, son of Speaker of House of Commons, and first cousin to the Lord Chancellor Finch of previous years, had a son who 'almost rivalled his father,' and who was made Solicitor-General and Earl of Aylesford.
Macclesfield: son, President of Royal Society.
Talbot: father was bishop, consecutively, of Oxford, Salisbury, and Durham; had sons, of one of whom there were great hopes, but he died young; the other 'succeeded to his father's virtues.'

Hardwick had five sons, all very distinguished. One, a man of letters; second, Lord Chancellor Yorke; third, an ambassador; fourth, 'talented as the others;' fifth, Bishop of Ely.

Northington: father was 'one of the most accomplished men of his day.'

Pratt: father was Chief Justice of King's Bench; his son was distinguished for public service.

Yorke: father was Lord Chancellor Hardwicke. (*See* above.)

Bathurst: father was the Lord Bathurst of Queen Anne's time; his son was the Lord Bathurst who filled high office under George III and IV.

Erskine: his brothers were nearly as eminent. The whole family was most talented.

Eldon: brother was the famous Lord Stowell, Judge of Admiralty.

Thus out of the 39 Chancellors 16 had kinsmen of eminence. 13 of them — viz. Sir Nicholas Bacon, Lord Bacon, Coventry, Littleton, Whitelock, Herbert, Finch, Hardwick, Pratt, Yorke, Bathurst, Erskine, and Eldon — had kinsmen of great eminence. In other words, 13 out of 39 — that is, 1 in every 3 — are remarkable instances of hereditary influence.

It is astonishing to remark the number of the Chancellors, who rose from mediocre social positions, showing how talent makes its way at the Bar, and how utterly insufficient are favouritism and special opportunities to win the great legal prize of the Chancellorship. It is not possible accurately, and it is hardly worth while roughly, to calculate the numerical value of hereditary influence in obtaining the Chancellorship. It is sufficient to say that it is enormous. We must not only reckon the number of students actually at the Chancery bar, and say that the Lord Chancellor was the foremost man among them, but we must reckon the immense number of schools in England, in any one of which, if a boy shows real marks of eminence, he is pretty sure to be patronised and passed on to a better place of education; whence by exhibitions, and subsequently by University scholarships and fellowships, he may become educated as a lawyer. I believe, from these reasons, that the chances of the son of a Lord Chancellor to be himself also a Chancellor, supposing he enters the law, to be more than a thousandfold greater than if he were the son of equally rich but otherwise undistinguished parents. It does not appear an accident that, out of 54 Lord Chancellors or Lord Keepers, two — viz. Sir Nicholas Bacon and Lord Hardwick — should have had sons who were also Chancellors, when we bear in mind the very eminent legal relationships of Herbert, Finch, Eldon, and the rest.

The intellectual force of English boys has, up to almost the present date, been steadily directed to classical education. Classics form the basis of instruction at our grammar schools, so that every boy who possesses signal classical aptitudes has a chance of showing them. Those who are successful obtain exhibitions and other help, and ultimately find their way to the great arena of competition of University life.

The senior classic at Cambridge is not only the foremost of the 300 youths who take their degrees in the same year, but he is the foremost of perhaps a tenth part of the classical intellect of his generation, throughout all England. No industry, without eminent natural talent to back it, could possibly raise a youth into that position.

The institution of the class list at Cambridge dates from 1824; so there have been 41 senior classics up to the present year. Wherever two names had been bracketed together, I selected the one that stood best in other examinations, and then extracted the following names from the list of them, as instances of hereditary influence:—

1827. Kennedy: father was a classic of eminence; two brothers, *see* below; another brother, almost equally distinguished in classics.
1828. Selwyn: brother M.P. for Cambridge, an eminent lawyer.
1830. Wordsworth nephew to the poet, brother of an almost equally distinguished classic, son of the Master of Trinity.
1831. Kennedy (*see* above).
1832. Lushington: brother (*see* below); nephew to the Right Hon. Sir Stephen Lushington. The family has numerous other members of eminent talent.
1834. Kennedy (*see* above).
1835. Goulbourn: father, Chancellor of Exchequer, nephew of Serjeant Goulbourn, cousin to Dr Goulbourn, Head Master of Rugby, the well-known preacher.
1835. Vaughan: many relationships like those of Goulbourn, including the Judge, the Professor at Oxford, and Mr Hawkins (*see* below).
1842. Denman: father was the eminent Chief Justice Lord Denman.
1846. Lushington: brother (*see* above).
1854. Hawkins: *see* Vaughan.

1855. Butler: son of Senior Wrangler of 1794; three brothers, of whom two held University Scholarships in Oxford, and the other was a double first-class man at Cambridge.

12 of the 41, or about 1 in three and a half, show these influences in a more or less marked degree; 7 of them, or 1 in 6, viz. 3 Kennedy, 1 Wordsworth, 2 Lushington, and 1 Butler, very much so.

The data we have been considering are summed up in the following table: —[3]

Number of cases.		Occurrence of near male relationship.	Percentages	
			Distinguished father has a distinguished son.	Distinguished man has distinguished brother.
605	All the men of "original minds" (Sir T. Phillips) and of every profession between 1453 and 1853	1 in 6 cases.	6 times in 100 cases.	2 times in 100 cases.
85	Living notabilities (Walford's "Men of the Times," letter A)	1 in 3.5 cases.	7 times in 100 cases.	7 times in 100 cases.
391	Painters of all dates (Bryan's Diety. A)	1 in 6 cases.	6 in 100 cases.	4 in 100 cases.
515	Musicians (Fétis Diety. A)	1 in 10 cases.	6 in 100 cases.	3 in 100 cases.
54	Lord Chancellors (Lord Campbell)	1 in 3 cases.	16 in 100 cases.	4 in 100 cases.
41	Senior Classics of Cambridge	1 in 4 cases.	Too recent.	10 in 100 cases.
	Averages	1 in 6 cases.	8 in 100 cases.	5 in 100 cases.

Everywhere is the enormous power of hereditary influence forced on our attention. If we take a list of the most brilliant standard writers of the last few years, we shall find a large share of the number have distinguished relationships. It would be difficult to set off, against the following instances, the same number

---

[3] Editor: A facsimile of this chart is appended to the end of this edition.

of names of men of equal eminence, whose immediate relatives were undistinguished. Bronté (Jane Eyre and her two sisters); Bulwer (and his brother the ambassador); Disraeli (father, author of "Curiosities of Literature"); Hallam (son, the subject of "In Memoriam"); Kingsley (two brothers eminent novelists, two others no less talented); Lord Macaulay (son of Zachary Macaulay); Miss Martineau (and her brother); Merivale, Herman and Charles (brothers); Dean Stanley (father the bishop, and popular writer on birds); Thackeray (daughter, authoress of "Elizabeth"); Tennyson (brother also a poet); Mrs. Trollope (son, Anthony).

As we cannot doubt that the transmission of talent is as much through the side of the mother as through that of the father, how vastly would the offspring be improved, supposing distinguished women to be commonly married to distinguished men, generation after generation, their qualities being in harmony and not in contrast, according to rules, of which we are now ignorant, but which a study of the subject would be sure to evolve!

It has been said by Bacon that "great men have no continuance." I, however, find that very great men are certainly not averse to the other sex, for some such have been noted for their illicit intercourses, and, I believe, for a corresponding amount of illegitimate issue. Great lawyers are especially to be blamed in this, even more than poets, artists, or great commanders. It seems natural to believe that a person who is not married, or who, if married, does not happen to have children, should feel himself more vacant to the attractions of a public or a literary career than if he had the domestic cares and interests of a family to attend to. Thus, if we take a list of the leaders in science of the present day, the small number of them who have families is very remarkable. Perhaps the best selection of names we can make, is from those who have filled the annual scientific office of President of the British Association. We will taken the list of the commoners simply, lest it should be objected, though unjustly, that some of the noblemen who have occupied the chair were not wholly indebted to their scientific attainments for that high position. Out of twenty-two individuals, about one-third have children; one-third are or have been married and have no children; and one-third have never been married. Among the children of those who have had families, the names of Frank Buckland and Alexander Herschel are already well-known to the public.

There has been a popular belief that men of great intellectual eminence, are usually of feeble constitution, and of a dry and cold disposition. There may be such instances, but I believe the general rule to be exactly the opposite. Such men, so far as by observation and reading extend, are usually more manly and genial than the average, and by the aid of these very qualities, they obtain a recognised ascendancy. It is a great and common mistake to suppose that high intellectual powers are commonly associated with puny frames and small physical strength. Men of remarkable eminence are almost always men of vast powers of work. Those among them that have fallen into sedentary ways will frequently astonish their friends by their physical feats, when they happen to be in the mood of a vacation ramble. The Alpine Club contains a remarkable number of men of fair literary and scientific distinction; and these are among the strongest and most daring of the climbers. I believe, from my own recollections of the thews and energies of my contemporaries and friends of many years at Cambridge, that the first half-dozen class-men in classics or mathematics would have beaten, out of all proportion, the last half-dozen class-men in any trial of physical strength or endurance. Most notabilities have been great eaters and excellent digesters, on literally the same principle that the furnace which can raise more steam than is usual for one of its size must burn more freely and well than is common. Most great men are vigorous animals, with exuberant powers, and an extreme devotion to a cause. There is no reason to suppose that, in breeding for the highest order of intellect, we should produce a sterile or a feeble race.

Many forms of civilization have been peculiarly unfavourable to the hereditary transmission of rare talent. None of them were more prejudicial to it than that of the Middle Ages, where almost every youth of genius was attracted into the Church, and enrolled in the ranks of a celibate clergy.

Another great hindrance to it is a costly tone of society, like that of our own, where it becomes a folly for a rising man to encumber himself with domestic expenses, which custom exacts, and which are larger than his resources are able to meet. Here also genius is celibate, at least during the best period of manhood.

A spirit of caste is also bad, which compels a man

of genius to select his wife from a narrow neighbourhood, or from the members of a few families.

But a spirit of clique is not bad. I understand that in Germany it is very much the custom for professors to marry the daughters of other professors, and I have some reason to believe, but am anxious for further information before I can feel sure of it, that the enormous intellectual digestion of German literary men, which far exceeds that of the corresponding class of our own countrymen, may, in some considerable degree, be traceable to this practice.

So far as beauty is concerned, the custom of many countries, of the nobility purchasing the handsomest girls they could find for their wives, has laid the foundation of a higher type of features among the ruling classes. It is not so very long ago in England that it was thought quite natural that the strongest lance at the tournament should win the fairest or the noblest lady. The lady was the prize to be tilted for. She rarely objected to the arrangement, because her vanity was gratified by the *éclat* of the proceeding. Now history is justly charged with a tendency to repeat itself. We may, therefore, reasonably look forward to the possibility, I do not venture to say the probability, of a recurrence of some such practice of competition. What an extraordinary effect might be produced on our race, if its object was to unite in marriage those who possessed the finest and most suitable natures, mental, moral, and physical!

Let us, then, give reins to our fancy, and imagine a Utopia — or a Laputa, if you will — in which a system of competitive examination for girls, as well as for youths, had been so developed as to embrace every important quality of mind and body, and where a considerable sum was yearly allotted to the endowment of such marriages as promised to yield children who would grow into eminent servants of the State. We may picture to ourselves an annual ceremony in that Utopia or Laputa, in which the Senior Trustee of the Endowment Fund would address ten deeply-blushing young men, all of twenty-five years old, in the following terms: —

> Gentlemen, I have to announce the results of a public examination, conducted on established principles; which show that you occupy the foremost places in your year, in respect to those qualities of talent, character, and bodily vigour which are proved, on the whole, to do most honour and best service to our race. An examination has also been conducted on established principles among all the young ladies of this country who are now of the age of twenty-one, and I need hardly remind you, that this examination takes note of grace, beauty, health, good temper, accomplished housewifery, and disengaged affections, in addition to noble qualities of heart and brain. By a careful investigation of the marks you have severally obtained, and a comparison of them, always on established principles, with those obtained by the most distinguished among the young ladies, we have been enabled to select ten of their names with

especial reference to your individual qualities. It appears that marriages between you and these ten ladies, according to the list I hold in my hand, would offer the probability of unusual happiness to yourselves, and, what is of paramount interest to the State, would probably result in an extraordinarily talented issue. Under these circumstances, if any or all of these marriages should be agreed upon, the Sovereign herself will give away the brides, at a high and solemn festival, six months hence, in Westminster Abbey. We, on our part, are prepared, in each case, to assign 5000£ as a wedding-present, and to defray the cost of maintaining and educating your children, out of the ample funds entrusted to our disposal by the State.

If a twentieth part of the cost and pains were spent in measures for the improvement of the human race that is spent on the improvement of the breed of horses and cattle, what a galaxy of genius might we not create! We might introduce prophets and high priests of civilization into the world, as surely as we can propagate idiots by mating *crétins*. Men and women of the present day are, to those we might hope to bring into existence, what the pariah dogs of the streets of an Eastern town are to our own highly-bred varieties.

The feeble nations of the world are necessarily giving way before the nobler varieties of mankind; and even the best of these, so far as we know them, seem unequal to their work. The average culture of mankind is become so much higher than it was, and the branches of knowledge and history so various and extended, that few are capable even of comprehending the exigencies of our modern civilization; much less of fulfilling them. We are living in a sort of intellectual anarchy, for the want of master minds. The general intellectual capacity of our leaders requires to be raised, and also to be differentiated. We want abler commanders, statesmen, thinkers, inventors, and artists. The natural qualifications of our race are no greater than they used to be in semi-barbarous times, though the conditions amid which we are born are vastly more complex than of old. The foremost minds of the present day seem to stagger and half under an intellectual load too heavy for their powers.

Part II[4]

I have shown, in my previous paper,[5] that intellectual capacity is so largely transmitted by descent that, out of every hundred sons of men distinguished in the open professions, no less than eight are found to have rivalled their fathers in eminence. It must be recollected that success of this kind implies the simultaneous inheritance of many points of character, in addition to mere intellectual capacity. A man must inherit good health, a love of mental work, a strong purpose, and considerable ambition, in order to achieve successes of the high order of which we are speaking. The deficiency of any one of these qualities would certainly be injurious, and probably be fatal to his chance of obtaining great distinction. But more than this: the proportion we have arrived at takes no account whatever of one-half of the hereditary influences that form the nature of the child. My particular method of inquiry did not admit of regard being paid to the influences transmitted by the mother, whether they had strengthened or weakened those transmitted by the father. Lastly, though the talent and character of both of the parents might, in any particular case, be of a remarkably noble order, and thoroughly congenial, yet they would necessarily have such mongrel antecedents that it would be absurd to expect their

---

[4] *MacMillan's Magazine*, vol. 12, 1865 pp. 318-327.
[5] Galton's Note — I take this opportunity of correcting a small *erratum* in my last paper. The name of the author of the forthcoming Brief Biographical Dictionary from which I quoted should have been the Rev. Charles Hole, not Hone.)

children to invariably equal them in their natural endowments. The law of atavism prevents it. When we estimate at its true importance this accumulation of impediments in the way of the son of a distinguished father rivalling his parent — the mother being selected, as it were, at haphazard — we cannot but feel amazed at the number of instances in which a successful rivalship has occurred. Eight per cent is as large a proportion as could have been expected on the most stringent hypothesis of hereditary transmission. No one, I think, can doubt, from the facts and analogies I have brought forward, that, if talented men were mated with talented women, of the same mental and physical characters as themselves, generation after generation, we might produce a highly-bred human race, with no more tendency to revert to meaner ancestral types than is shown by our long-established breeds of race-horses and fox-hounds.

It may be said that, even granting the validity of my arguments, it would be impossible to carry their indications into practical effect. For instance, if we divided the rising generation into two castes, A and B, of which A was selected for natural gifts, and B was the refuse, then, supposing marriage was confined within the pale of the caste to which each individual belonged, it might be objected that we should simply differentiate our race — that we should create a good and a bad caste, but we should not improve the race as a whole. I reply that this is by no means the necessary result. There remains another very important law to be brought into play. Any agency, however indirect, that would somewhat hasten the marriages in caste A, and retard those in caste B, would result in a larger proportion of children being born to A than to B, and would end by wholly eliminating B, and replacing it by A.

Let us take a definite case, in order to give precision to our ideas. We will suppose the population to be, in the first instance, stationary; A and B to be equal in numbers; and the children of each married pair who survive to maturity to be rather more than two and a half in the case of A, and rather less than one and a half in the case of B. This no extravagant hypothesis. Half the population of the British Isles are born of mothers under the age of thirty years.

The result in the first generation would be that the total population would be unchanged, but that only one-third part of it would consist of the children of B. In the second generation, the descendants of B would be reduced to two-ninths of their original numbers, but the total population would begin to increase, owing to the greater preponderance of the prolific caste A. At this point the law of natural selection would powerfully assist in the substitution of caste A for caste B, by pressing heavily on the minority of weakly and incapable men.

The customs that affect the direction and date of marriages are already numerous. In many families, marriages between cousins are discouraged and checked. Marriages, in other respects appropriate, are very commonly deferred, through prudential considerations. If it was generally felt that intermarriages between A and B were as unadvisable as they are supposed to be between cousins, and that marriages in A ought to be hastened, on the ground of prudential considerations, while those in B ought to be discouraged and retarded, then, I believe, we should have agencies amply sufficient to eliminate B in a few generations.

I hence conclude that the improvement of the breed of mankind is no insuperable difficulty. If everybody were to agree on the improvement of the race of man being a matter of the very utmost importance, and if the theory of the hereditary transmission of qualities in men was as thoroughly understood as it is in the case of our domestic animals, I see no absurdity in supposing that, in some way or other, the improvement would be carried into effect.

It remains for me in the present article to show that hereditary influence is as clearly marked in mental aptitudes as in general intellectual power. I will then enter into some of the considerations which my views on hereditary talent and character naturally suggest.

I will first quote a few of those cases in which characteristics have been inherited that clearly depend on peculiarities of organization. Prosper Lucas was among our earliest encyclopædists on this subject. It is distinctly shown by him, and agreed to by others, such as Mr. G. Lewes, that predisposition to any form of disease, or any malformation, may become an inheritance. Thus disease of the heart is hereditary; so are tubercles in the lungs; so also are diseases of the brain, of the liver, and of the kidney; so are diseases of the eye and of the ear. General maladies are equally inheritable, as gout and madness. Longevity on the one hand, and premature deaths on the other, go by descent. If we consider a class of peculiarities, more recondite in their origin than these, we shall still find the law of inheritance to hold good. A morbid susceptibility to contagious disease, or to the poisonous effects of opium, or of calomel, and an aversion to the taste of meat, are all found to be inherited. So is a craving for drink, or for gambling, strong sexual passion, a proclivity to pauperism, to crimes of violence, and to crimes of fraud.

There are certain marked types of character, justly associated with marked types of feature and of temperament. We hold, axiomatically, that the latter are inherited (the case being too notorious, and too consistent with the analogy afforded by brute animals, to render argument necessary), and we therefore infer the same of the former. For instance, the face of the combatant is square, course, and heavily jawed. It differs from that of the ascetic, the voluptuary, the dreamer, and the charlatan.

Still more strongly marked than these, are the typical features and characters of different races of men. The Mongolians, Jews, Negroes, Gipsies, and American Indians; severally propagate their kinds; and each kind differs in character and intellect, as well as in colour and shape, from the other four. They, and a vast number of other races, form a class of instances worthy of close investigation, in which peculiarities of character are invariably transmitted from the parents to the offspring.

In founding argument on the innate character of different races, it is necessary to bear in mind the exceeding docility of man. His mental habits in mature life are the creatures of social discipline, as well as of inborn aptitudes, and it is impossible to ascertain what is due to the latter alone, except by observing several individuals of the same race, reared under various influences, and noting the peculiarities of character that invariably assert themselves. But, even when we have imposed these restrictions to check a hasty and imaginative conclusion, we find there remain abundant data to prove an astonishing diversity in the natural characteristics of different races. It will be sufficient for our purpose of we fix our attention upon the peculiarities of one or two of them.

The race of the American Indians is spread over an enormous area, and through every climate; for it reaches from the frozen regions of the North, through the equator down to the inclement regions of the South. It exists in thousands of disconnected communities, speaking nearly as many different languages. It has been subjected to a strange variety of political influences, such as its own despotisms in Peru, Mexico, Natchez, and Bogota, and its numerous republics, large and small. Members of the race have been conquered and ruled by military adventures from Spain and Portugal; others have been subjugated to Jesuitical rule; numerous settlements have been made by strangers on its soil; and, finally, and north of the continent has been colonized by European races. Excellent observers have watched the American Indians under all these influences, and their almost unanimous conclusion is as follows:—

The race is divided into many varieties, but it has fundamentally the same character throughout the whole of America. The men, and in a less degree the women, are naturally cold, melancholic, patient, and taciturn. A father, mother, and their children, are said to live together in a hut, like persons assembled by accident, not tied by affection. The youths treat their parents with neglect, and often with such harshness and insolence as to horrify Europeans who have witnessed their conduct. The mothers have been seen to commit infanticide without the slightest discomposure, and numerous savage tribes have died out in consequence of this practice. The American Indians are eminently non-gregarious. They nourish a sullen reserve, and show little sympathy with each other, even when in great distress. The Spaniards had to enforce the common duties of humanities by positive laws. They are strangely taciturn. When not engaged in action they will sit whole days in one posture without opening their lips, and wrapped up in their narrow thoughts. The usually march in Indian file, that is to say, in a long line, at some distance from each other, without exchanging a word. They keep the same profound silence in rowing a canoe, unless they happen to be excited by some extraneous cause. On the other hand, their patriotism and local attachments are strong, and they have an astonishing sense of personal dignity. The nature of the American Indians appears to contain the minimum of affectionate and social qualities compatible with the continuance of their race.

Here, then, is a well-marked type of character, that formerly prevailed over a large part of the globe, with which other equally marked types of character in other regions are strongly contrasted. Take, for instance, the typical West African Negro. He is more unlike the Red man in his mind than in his body. Their characters are almost opposite, one to the other. The Red man has great patience, great reticence, great dignity, and no passion; the Negro has strong impulsive passions, and neither patience, reticence, nor dignity. He is warm-hearted, loving towards his master's children, and idolised by the children in return. He is eminently gregarious, for he is always jabbering, quarrelling, tom-tom-ing, or dancing. He is remarkably domestic, and he is endowed with such constitutional vigour, and is so prolific, that his race is irrepressible.

The Hindu, the Arab, the Mongol, the Teuton, and very many more, have each of them their peculiar characters. We have not space to analyse them on this occasion; but, whatever they are, they are transmitted, generation after generation, as truly as their physical forms.

What is true for the entire race is equally true for its varieties. If we were to select persons who were born with a type of character that we desired to intensify, — suppose it was one that approached to some ideal standard of perfection — and if we compelled marriage within the limits of the society so selected, generation after generation; there can be no doubt that the offspring would ultimately be born with the qualities we sought, as surely as if we had been breeding for physical features, and not for intellect or disposition.

Our natural constitution seems to bear as direct and stringent a relation to that of our forefathers as any other physical effect does to its cause. Our bodies, minds, and capabilities of development have been derived from them. Everything we possess at our birth is a heritage from our ancestors.

Can we hand anything down to our children, that

we have fairly won by our own independent exertions? Will our children be born with more virtuous dispositions, if we ourselves have acquired virtuous habits? Or are we no more than passive transmitters of a nature we have received, and which we have no power to modify? There are but a few instances in which habit even seems to be inherited. The chief among them are such as those of dogs being born excellent pointers; of the attachment to man shown by dogs; and of the fear of man, rapidly learnt and established among the birds of newly-discovered islands. But all of these admit of being accounted for on other grounds than the hereditary transmission of habits. Pointing is, in some faint degree, a natural disposition of all dogs. Breeders have gradually improved upon it, and created the race we now possess. There is nothing to show that the reason why dogs are born staunch pointers if that their parents had been broken into acquiring an artificial habit. So as regards the fondness of dogs for man. It is inherent to a great extent in the genus. The dingo, or wild dog of Australia, is attached to the man who has caught him when a puppy, and clings to him even although he is turned adrift to hunt for his own living. This quality in dogs is made more intense by the custom of selection. The savage dogs are lost or killed; the tame ones are kept and bred from. Lastly, as regards the birds. As soon as any of their flock has learned to fear, I presume that its frightened movements on the approach of man form a language that is rapidly and unerringly understood by the rest, old or young; and that, after a few repetitions of the signal, man becomes an object of well-remembered mistrust.

Moreover, just as natural selection has been shown to encourage love of man in domestic dogs, so it tends to encourage fear of man in all wild animals — the tamer varieties perishing owing to their misplaced confidence, and the wilder ones continuing their breed.

If we examine the question from the opposite side, a list of life-long habits in the parents might be adduced which leave no perceptible trace on their descendants. I cannot ascertain that the son of an old soldier learns his drill more quickly than the son of an artizan. I am assured that the sons of fishermen, whose ancestors have pursued the same calling time out of mind, are just as sea-sick as the sons of landsmen when they first go to sea. I cannot discover that the castes of India show signs of being naturally endowed with special aptitudes. If the habits of an individual are transmitted to his descendants, it is, as Darwin says, in a very small degree, and is hardly, if at all, traceable.

We shall therefore take an approximately correct view of the origin of our life, if we consider our own embryos to have sprung immediately from those embryos whence our parents were developed, and these from the embryos of *their* parents, and so on for ever. We should in this way look on the nature of mankind, and perhaps on that of the whole animated creation, as one continuous system, ever pushing out new branches in all directions, that variously interlace, and that bud into separate lives at every point of interlacement.

This simile does not at all express the popular notion of life. Most persons seem to have a vague idea that a new element, specially fashioned in heaven, and not transmitted by simple descent, is introduced into the body of every newly-born infant. Such a notion is unfitted to stand upon any scientific basis with which we are acquainted. It is impossible it should be true, unless there exists some property or quality in man that is not transmissible by descent. But the terms *talent* and *character* are exhaustive: they include the whole of man's spiritual nature so far as we are able to understand it. No other class of qualities is known to exist, that we might suppose to have been interpolated from on high. Moreover, the idea is improbable from *à priori* considerations, because there is no other instance in which creative power operates under our own observation at the present day, except it may be in the freedom in action of our own wills. Wherever else we turn our eyes, we see nothing but law and order, and effect following cause.

But though, when we look back to our ancestors, the embryos of our progenitors may be conceived to have been developed, in each generation, immediately from the one that preceded it, yet we cannot take so restricted a view when we look forward. The interval that separates the full-grown animal from its embryo is too important to be disregarded. It is in this interval that Darwin's law of natural selection comes into play; and those conditions are entered into, which affect, we know not how, the 'individual variation' of the offspring. I mean those that cause dissimilarity among brothers and sisters who are born successively, while twins, produced simultaneously, are often almost identical. If it were possible that embryos should descent directly from embryos, there might be developments in every direction, and the world would be filled with monstrosities. But this is not the order of nature. It is her fiat that the natural tendencies of animals should never disaccord long and widely with the conditions under which they are placed. Every animal before it is of an age to bear offspring, has to undergo frequent stern examinations before the board of nature, under the laws of natural selection; where to be 'plucked' is not necessarily disgrace, but is certainly death. Never let it be forgotten that man, as a reasonable being, has the privilege of not being helpless under the tyranny of uncongenial requirements, but that he can, and that he does, modify the subjects in which nature examines him, and that he has considerable power in settling beforehand the relative importance in the examination that shall be assigned to each separate subject.

It becomes a question of great interest how far moral monstrosities admit of being bred. Is there any obvious law that assigns a limit to the propagation of supremely vicious or supremely virtuous natures? In strength, agility, and other physical qualities, Darwin's law of natural selection acts with unimpassioned, merciless severity. The weakly die in the battle for life; the stronger and more capable individuals are alone permitted to survive, and to bequeath their constitutional vigour to future generations. Is there any corresponding rule in respect to moral character? I believe there is, and I have already hinted at it when speaking of the American Indians. I am prepared to maintain that its action, by insuring a certain fundamental unity in the quality of the affections, enables men and the higher order of animals to sympathise in some degree with each other, and also, that this law forms the broad basis of our religious sentiments.

Animal life, in all but the very lowest classes, depends on at least one, and, more commonly, on all of the four following principles: — There must be affection, and it must be of four kinds: sexual, parental, filial, and social. The absolute deficiency of any one of these would be a serious hindrance, if not a bar to the continuance of any race. Those who possessed all of them, in the strongest measure, would, speaking generally, have an advantage in the struggle for existence. Without sexual affection, there would be no marriages, and no children; without parental affection, the children would be abandoned; without filial affection, they would stray and perish; and, without the social, each individual would be single-handed against rivals who were capable of banding themselves into tribes. Affection for others as well as self, is therefore a necessary part of animal character. Disinterestedness is as essential to a brute's well-being as selfishness. No animal lives for itself alone, but also, at least occasionally, for its parent, its mate, its offspring, or its fellow. Companionship is frequently more grateful to an animal than abundant food. The safety of her young is considered by many a mother as a paramount object to her own. The passion for a mate is equally strong. The gregarious bird posts itself during its turn of duty as watchman on a tree, by the side of the feeding flock. Its zeal to serve the common cause exceeds its care to attend to its own interests. Extreme selfishness is not a common vice. Narrow thoughts of self by no means absorb the minds of ordinary men; they occupy a secondary position in the thoughts of the more noble and generous of our race. A large part of an Englishman's life is devoted to others, or to the furtherance of general ideas, and not to directly personal ends. The Jesuit toils for his order, not for himself. Many plan for that which they can never live to see. At the hour of death they are still planning. An in-completed will, which might work unfairness among those who would succeed to the property of a dying man, harasses his mind. Personal obligations of all sorts press as heavily as in the fullness of health, although the touch of death is known to be on the point of cancelling them. It is so with animals. A dog's thoughts are towards his master, even when he suffers the extremest pain. His mind is largely filled at all times with sentiments of affection. But disinterested feelings are more necessary to man than to any other animal, because of the long period of his dependent childhood, and also because of his great social needs, due to his physical helplessness. Darwin's law of natural selection would therefore be expected to develop these sentiments among men, even among the lowest barbarians, to a greater degree than among animals.

I believe that our religious sentiments spring primarily from these four sources. The institution of celibacy is an open acknowledgment that the theistic and human affections are more or less convertible; I mean that by starving the one class the other becomes more intense and absorbing. In savages, the theistic sentiment is chiefly, if not wholly, absent. I would refer my readers, who may hesitate in accepting this assertion, to the recently published work of my friend

Sir John Lubbock, "Prehistoric Times," p. 467-472, where the reports of travellers on the religion of savages are very ably and fairly collated. The theistic sentiment is secondary, not primary. It becomes developed within us under the influence of reflection and reason. All evidence tends to show that man is directed to the contemplation and love of God by instincts that he shares with the whole animal world, and that primarily appeal to the love of his neighbour.

Moral monsters are born among Englishmen, even at the present day; and, when they are betrayed by their acts, the law puts them out of the way, by the prison or the gallows, and so prevents them from continuing their breed. Townley, the murderer, is an instance in point. He behaved with decorum and propriety; he was perfectly well-conducted to the gaol officials, and he corresponded with his mother in a style that was certainly flippant, but was not generally considered to be insane. However, with all this reasonableness of disposition, he could not be brought to see that he had done anything particularly wrong in murdering the girl that was disinclined to marry him. He was thoroughly consistent in his disregard for life, because, when his own existence became wearisome, he ended it with perfect coolness, by jumping from an upper staircase. It is a notable fact that a man without a conscience, like Townley, should be able to mix in English society for years, just like other people.

How enormous is the compass of the scale of human character, which reaches from dispositions like those we have just described, to that of Socrates! How various are the intermediate types of character that commonly fall under everybody's notice, and how differently are the principles of virtue measured out to different natures! We can clearly observe the extreme diversity of character in children. Some are naturally generous and open, others mean and tricky; some are warm and loving, others cold and heartless; some are meek and patient, others obstinate and self-asserting; some few have the tempers of angels, and at least as many have the tempers of devils. In the same way, as I showed in my previous paper, that by selecting men and women of rare and similar talent, and mating them together, generation after generation, an extraordinarily gifted race might be developed, so a yet more rigid selection, having regard to their moral nature, would, I believe, result in a no less marked improvement of their natural disposition.

 Let us consider an instance in which different social influences have modified the inborn dispositions of a nation. The North American people has been bred from the most restless and combative class of Europe. Whenever, during the last ten or twelve generations, a political or religious party has suffered defeat, its prominent members, whether they were the best, or only the noisiest, have been apt to emigrate to America, as a refuge from persecution. Men fled to America for conscience sake, and for that of unappreciated patriotism. Every scheming knave, and every brutal ruffian, who feared the arm of the law, also turned his eyes in the same direction. Peasants and artisans, whose spirit rebelled against the tyranny of society and the monotony of their daily life, and men of a higher position, who chafed under conventional restraints, all yearned towards America. Thus the dispositions of the parents of the American people have been exceedingly varied, and usually extreme, either for good or for evil. But in one respect they almost universally agreed. Every head of an emigrant family brought with him a restless character, and a spirit apt to rebel. If we estimate the moral nature of Americans from their present social state, we shall find it to be just what we might have expected from such a parentage. They are enterprising, defiant, and touchy; impatient of authority; furious politicians; very tolerant of fraud and violence; possessing much high and generous spirit, and some true religious feeling, but strongly addicted to cant.

We have seen that the law of natural selection develops disinterested affection of a varied character even in animals and barbarian man. Is the same law different in its requirements when acting on civilized man? It is no doubt more favourable on the whole to civilized progress, but we must not expect to find as yet many marked signs of its action. As a matter of history, our Anglo-Saxon civilization is only skin-deep. It is but eight hundred years, or twenty-six generations, since the Conquest, and the ancestors of the large majority of Englishmen were the merest boors at a much later date than that. It is said that among the heads of the noble houses of England there can barely be found one that has a right to claim the sixteen quarterings — that is to say, whose great-great-grandparents were, all of them (sixteen in number), entitled to carry arms. Generally the nobility of a family is represented by only a few slender rills among a multiplicity of non-noble sources.

The most noble quality that the requirements of

civilization have hitherto bred in us, living as we do in a rigorous climate and on a naturally barren soil, is the instinct of continuous steady labour. This is alone possessed by civilized races, and it is possessed in a far greater degree by the feeblest individuals among them than by the most able-bodied savages. Unless a man can work hard and regularly in England, he becomes an outcast. If he only works by fits and starts he has not a chance of competition with steady workmen. An artizan who has variable impulses, and wayward moods, is almost sure to end in intemperance and ruin. In short, men who are born with wild and irregular dispositions, even though they contain much that is truly noble, are alien to the spirit of a civilized country, and they and their breed are eliminated from it by the law of selection. On the other hand, a wild, untameable restlessness is innate with savages. I have collected numerous instances where children of a low race have been separated at an early age from their parents, and reared as part of a settler's family, quite apart from their own people. Yet, after years of civilized ways, in some fit of passion, or under some craving, like that of a bird about to emigrate, they have abandoned their home, flung away their dress, and sought their countrymen in the bush, among whom they have subsequently been found living in contented barbarism, without a vestige of their gentle nurture. This is eminently the case with the Australians, and I have heard of many others in South Africa. There are also numerous instances in England where the restless nature of gipsy half-blood asserts itself with irresistible force.

Another different, which may either be due to

natural selection or to original difference of race, is the face that savages seem incapable of progress after the first few years of their life. The average children of all races are much on a par. Occasionally, those of the lower races are more precocious than the Anglo-Saxons; as a brute beast of a few weeks old is certainly more apt and forward than a child of the same age. But, as the years go by, the higher races continue to progress, while the lower ones gradually stop. They remain children in mind, with the passions of grown men. Eminent genius commonly asserts itself in tender years, but it continues long to develop. The highest minds in the highest race seem to have been those who had the longest boyhood. It is not those who were little men in early youth who have succeeded. Here I may remark that, in the great mortality that besets the children of our poor, those who are members of precocious families, and who are therefore able to help in earning wages at a very early age, have a marked advantage over their competitors. They, on the whole, live, and breed their like, while the others die. But, if this sort of precocity be unfavourable to a race — if it be generally followed up by an early arrest of development, and by a premature old age — then modern industrial civilization, in encouraging precocious varieties of men, deteriorates the breed.

Besides these three points of difference — endurance of steady labour, tameness of disposition, and prolonged development — I know of none that very markedly distinguishes the nature of the lower classes of civilized man from that of barbarians. In the excitement of a pillaged town the English soldier is just as brutal as the savage. Gentle manners seem, under those circumstances, to have been a mere gloss thrown by education over a barbarous nature. One of the effects of civilization is to diminish the rigour of the application of the law of natural selection. It preserves weakly lives, that would have perished in barbarous lands. The sickly children of a wealthy family have a better chance of living and rearing offspring than the stalwart children of a poor one. As with the body, so with the mind. Poverty is more adverse to early marriages than is natural bad temper, or inferiority of intellect. In civilized society, money interposes her ægis between the law of natural selection and very many of its rightful victims. Scrofula and madness are naturalised among us by wealth; short-sightedness is becoming so. There seems no limit to the morbific tendencies of body or mind that might accumulate in a land where the law of primogeniture was general, and where riches were more esteemed than personal qualities. Neither is there any known limit to the intellectual and moral grandeur of nature that might be introduced into aristocratical families, if their representatives, who have such rare privilege in winning wives that please them best, should invariably, generation after generation, marry with a view of transmitting those noble qualities to their descendants. Inferior blood in the representative of a family might be eliminated from it in a few generations. The share that a man retains in the constitution of his remote descendants is inconceivably small. The father transmits, on an average, one-half of his nature, the grandfather one-fourth, the great-grandfather one-eighth; the share decreasing step by step, in a geometrical ratio, with great rapidity. Thus the man who claims descent from a Norman baron, who accompanied William the Conqueror twenty-six generations ago, has so minute a share of that baron's influence in his constitution, that, if he weighs fourteen stone, the part of him which may be ascribed to the baron (supposing, of course, there have been no additional lines of relationship) is only one-fiftieth of a grain in weight— an amount ludicrously disproportioned to the value popularly ascribed to ancient descent. As a stroke of policy, I question if the head of a great family, or a prince, would not give more strength to his position, by marrying a wife who would bear him talented sons than one who would merely bring him the support of high family connexions.

With the few but not insignificant exceptions we have specified above, we are still barbarians in our nature, and we show it in a thousand ways. The children who dabble and dig in the dirt have inherited the instincts of untold generations of barbarian forefathers, who dug with their nails for a large fraction of their lives. Our ancestors were grubbing by the hour, each day, to get at the roots they chiefly lived upon. They had to grub out pitfalls for their games, holes for their palisades and hut-poles, hiding-places, and ovens. Man became a digging animal by nature; and so we see the delicately-reared children of our era very ready to revert to primeval habits. Instinct breaks out in them, just as it does in the silk-haired, boudoir-nurtured spaniel, with a ribbon round its neck, that runs away from the endearments of its mistress, to sniff and revel in some road-side mess of carrion.

It is a common theme of moralists of many creeds, that man is born with an imperfect nature. He has lofty aspirations, but there is a weakness in his disposition that incapacitates him from carrying his nobler purposes into effect. He sees that some particular course of action is his duty, and should be his delight; but his inclinations are fickle and base, and do not conform to his better judgment. The whole moral nature of man is tainted with sin, which prevents him from doing the things he knows to be right.

I venture to offer an explanation of this apparent anomaly, which seems perfectly satisfactory from a scientific point of view. It is neither more nor less than that the development of our nature, under Darwin's law of natural selection, has not yet overtaken the development of our religious civilization. Man was barbarous but yesterday, and therefore it is not to be expected that the natural aptitudes of his race should already have become moulded into accordance with his very recent advance. We men of the present centuries are like animals suddenly transplanted among new conditions of climate and of food: our instincts fail us under the altered circumstances.

My theory is confirmed by the fact that the members of old civilizations are far less sensible than those newly converted from barbarism of their nature being inadequate to their moral needs. The conscience of a Negro is aghast at his own wild, impulsive nature, and is easily stirred by a preacher, but it is scarcely possible to ruffle the self-complacency of a steady-going Chinaman.

The sense of original sin would show, according to my theory, not that man was fallen from a high estate, but that he was rapidly rising from a low one. It would therefore confirm the conclusion that has been arrived at by every independent line of ethnological research — that our forefathers were utter savages from the beginning; and, that, after myriads of years of barbarism, our race has but very recently grown to be civilized and religious.

Opposite: facsimiles of the charts reproduced above.

Art.	Lit. & science.	More distant.	Brothers.	Father & son.	Number.	NOTABLE PERSONS.
–	–	1	–	2	3	J. Adams, Pres. U.S.A.; son Samuel also patriot; nephew, J. Quincey, president.
–	2	–	2	–	2	W. Belsham, historian; brother of T. Belsham, Unitarian minister.
–	3	–	–	2	3	J. Bernouilli, father of James and uncle of John, all mathematicians.
3	–	–	–	2	3	Breughel, father and two sons, painters.
–	2	–	–	2	2	Buxtorff, father and son, Hebraists.
3	–	1	2	–	3	Caracci, An. and Ag. brothers, Lud. cousin, painter.
–	1	–	2	–	2	Cartwright, reformer; brother, mechanist.
–	3	–	–	3	3	Casini, grandfather, father, and son, all mathematicians.
–	1	2	–	–	2	Cooper, Privy Councillor to Cromwell; grandson, literary.
–	–	–	2	–	2	De Witt, two brothers, patriots.
–	–	–	–	3	3	Elizabeth, queen, daughter of Henry VIII. and granddaughter of Sir T. Bullen.
–	2	–	2	–	2	Fontana, two brothers, natural philosophers.
–	2	–	–	2	2	Forster, father and son, naturalists (Cook's voyages).
–	6	–	–	6	6	Gronovius, sons and grandsons, six in all, learned critics.
–	–	1	–	2	3	Gustavus Adolphus, father of Christina and grandson of Gustavus Vasa.
–	2	–	–	2	2	Herschel, father and son, astronomers.
–	2	–	2	–	2	Hunter, two brothers, anatomists.
–	2	2	–	–	2	Jussieu, uncle and nephew, botanists.
–	–	–	–	4	4	Medici, grandfather, father, and son, and Catherine.
–	–	–	–	2	2	Orleans, Egalité, and son Louis Phillippe.
–	–	–	2	–	2	Ostade, two brothers, painters.
–	4	–	4	–	4	Perrault, four brothers, all writers.
–	1	–	–	2	2	Penn, admiral; son, Quaker writer.
–	–	2	–	2	4	Phillibert, Prince of Orange; cousin William, whose son was Maurice. His grandson was our William III.
–	–	–	–	2	2	Pitt, father and son, statesmen.
–	2	–	–	2	2	Scaliger, classical critic; son also.
–	–	–	–	2	2	Sforzas, father and son.
–	1	2	–	–	2	Shaftesbury, statesman; grandson, author.
–	1	–	–	2	2	Sheridan, father and son.
–	1	–	–	2	2	Staël, Madam, daughter of Necker, financier.
–	6	–	–	6	6	Stephens, family of six, critics and editors.
2	–	–	–	2	2	Teniers, father and son, painters.
–	–	–	–	2	2	Tytler, historian and poet; son, Lord Woodhouselee.
2	–	–	–	2	2	Vandervelde, father and son, painters.
–	2	–	2	–	2	Vanderwurf, two brothers, famous for small history.
–	–	1	2	–	3	Valnoo, two brothers, and nephew, painters.
–	1	–	–	2	2	Walpole, Sir Robert, statesman; Sir Horace, author.
–	–	–	–	2	2	Van Tromp, father and son, admirals.
–	1	2	–	–	2	Villiers, statesman; grandson, the reprobate poet.
–	2	–	–	2	2	Vossius, father, son, and other relatives, all writers.
–	2	–	–	2	2	Warton, editor of Pope; son, poet.
9	52	14	22	66	102	

Number of cases.		Occurrence of near male relationship.	Percentages.	
			Distinguished father has a distinguished son.	Distinguished man has a distinguished brother.
605	All the men of "original minds" (Sir T. Phillips), and of every profession between 1453 and 1853	1 in 6 cases.	6 times in 100 cases.	2 times in 100 cases.
85	Living notabilities (Walford's "Men of the Times," letter A)	1 in 3½ cases.	7 ,,	7 ,,
391	Painters of all dates (Bryan's Dicty. A)	1 in 6 cases.	5 ,,	4 ,,
515	Musicians (Fétis Dicty. A)	1 in 10 cases.	6 ,,	3 ,,
54	Lord Chancellors (Lord Campbell)	1 in 3 cases.	10 ,,	4 ,,
41	Senior Classics of Cambridge	1 in 4 cases.	Too recent	10 ,,
	Averages	1 in 6 cases.	8 in 100 cases.	5 in 100 cases.

www.ingramcontent.com/pod-product-compliance
Lightning Source LLC
Chambersburg PA
CBHW070828100426
42813CB00003B/528